Artemis Flies to the Rescue

Written by
Ingrid Alesich

Illustrated by
Sean Winburn

AuthorHouse™
1663 Liberty Drive
Bloomington, IN 47403
www.authorhouse.com
Phone: 1 (800) 839-8640

© 2015 Ingrid Alesich. All rights reserved.

No part of this book may be reproduced, stored in a retrieval system, or transmitted by any means without the written permission of the author.

Published by AuthorHouse 10/30/2015

ISBN: 978-1-5049-5795-3 (sc)
ISBN: 978-1-5049-5796-0 (e)

Library of Congress Control Number: 2015917737

Print information available on the last page.

Any people depicted in stock imagery provided by Thinkstock are models, and such images are being used for illustrative purposes only.
Certain stock imagery © Thinkstock.

This book is printed on acid-free paper.

Because of the dynamic nature of the Internet, any web addresses or links contained in this book may have changed since publication and may no longer be valid. The views expressed in this work are solely those of the author and do not necessarily reflect the views of the publisher, and the publisher hereby disclaims any responsibility for them.

authorHOUSE®

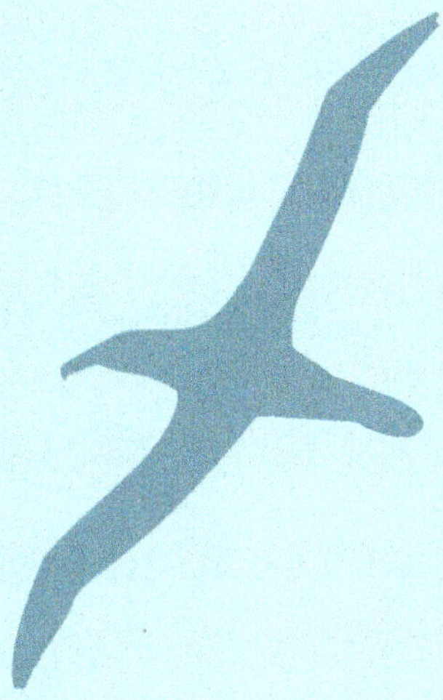

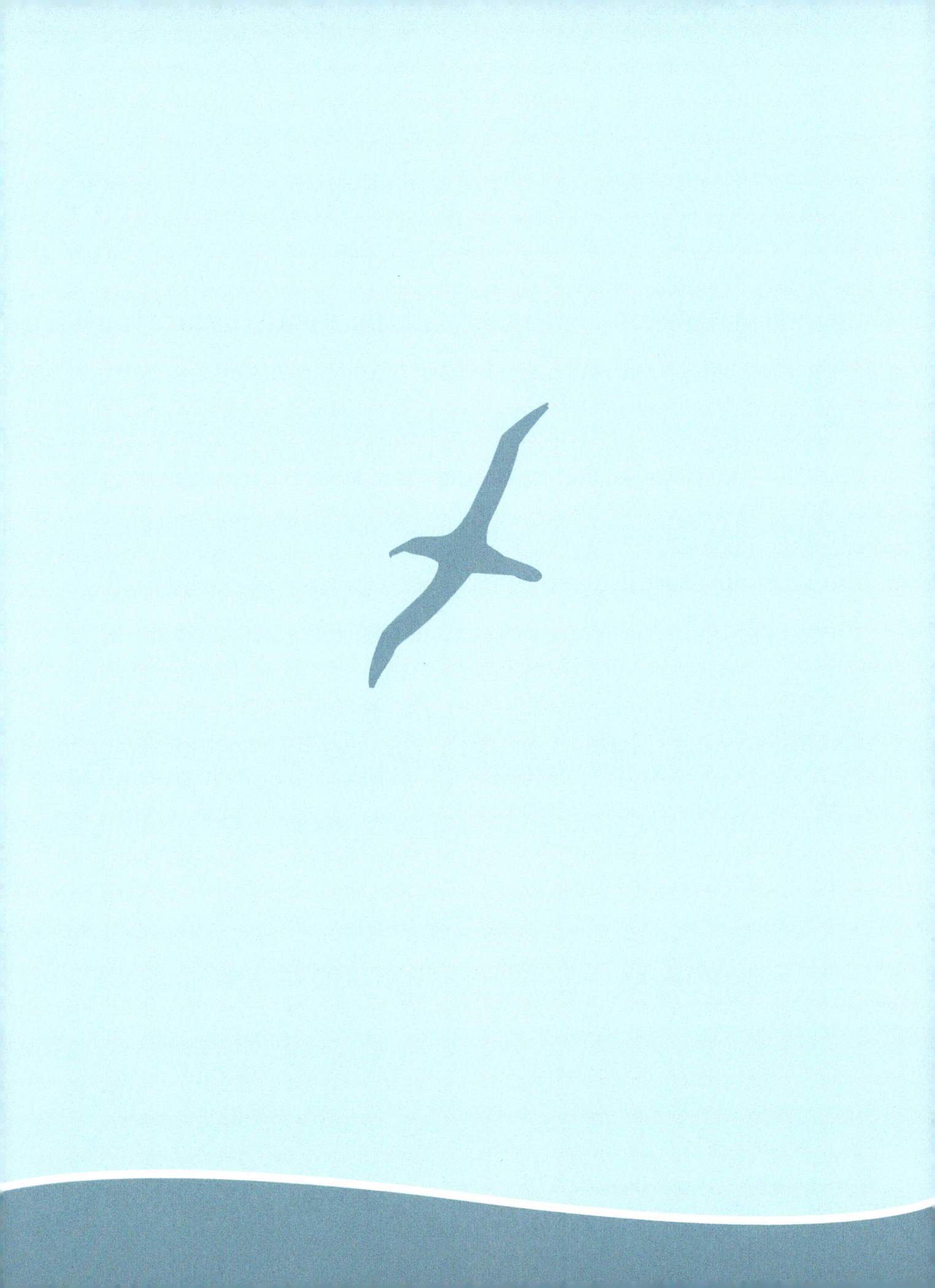

Artemis, the albatross, lives on a very big, blue ocean. Her wings are the widest wings in the world. From tip to tip, they can reach each end of a long sailboat, or from toe to head of 3 six year old children lying on the floor. Her strong chest and wings push against heavy ocean winds. She loves to fly far out and plunge deep into the waves catching an octopus, or a lobster, or a fish. She loves to take fish in her gullet back to her family. They live with hundreds of relatives on a lonely island with tall cliffs. Every morning, the adult albatrosses sing and dance together:

Earth my body Water my blood

Air my breath and Fire my spirit

We are one.

Then they flap their wings and take flight from the cliffs.

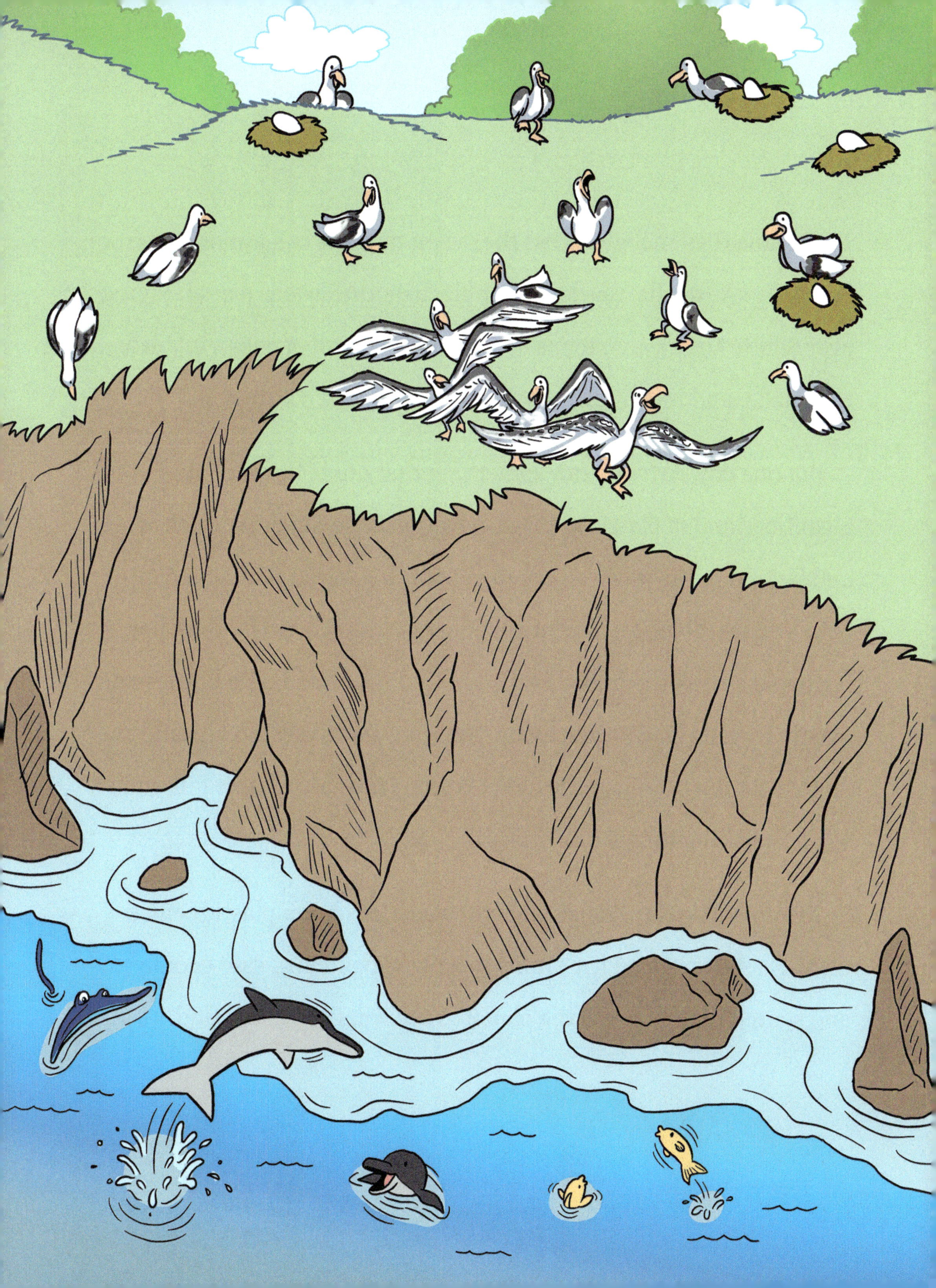

Artemis flies so high across the ocean that she can almost see from Canada to Australia. She loves her big, beautiful water planet. Flying over South America, Artemis sees rainforests full of colourful birds, waterfalls, and mountain meadows painted with wildflowers.

But one day, Artemis saw something that made her very angry. A gigantic island of plastic garbage, as big as Lake Superior, bobbed and heaved in the waves. Flapping down closer, she saw dead fish, baby whales, dolphins, one of her cousins, and stingrays. With her big hooked beak she pulled at the knotted nets filled with bottles and plastic wrap, trying to free them. The island was way too big, and she almost got tangled up in the nets herself. The plastic never falls apart, so animals caught in it die, trapped.

Artemis thought, I must find out where this killer plastic comes from. So she wrapped her family in a big winged hug, said goodbye, and soared high to travel the planet to see what she could find.

ARCTIC NORTH
ASIA
CANADA
U.S.
SOUTH AMERICA
AUSTRALIA
ANTARCTICA

First Artemis flew over Australia. In the cities, lots of smoke from big factories spewed black clouds into the sky. She saw more that shocked her. Inland, many forest fires burned. She coughed and covered her beak to protect herself from the choking smoke. The rivers had all dried up; the lakes and ponds had bottoms of hard earth, no water. Kangaroos were fleeing into the cities because their homes in the forests were burning. Plastic garbage was floating in the bays.

Artemis soared north to Asia and saw rivers flooding and thick, black, coal smoke everywhere. Hurricanes hurled below her, as she heaved her way west over to Africa. Rivers and lakes had dried there too, so lions, hippos, wildebeests as well as people were sick and starving. Their mouths were dry with thirst under the blazing, hot sun. Without water, no food could grow. Plastic garbage was piled up in Asia and Africa, in India.

This is awful!" she cried. "What happened to the rain? What happened to the beautiful forests and fields? Who is making all this plastic? Who is dumping the plastic garbage in the rivers, bays and oceans? What is causing this terrible pollution?"

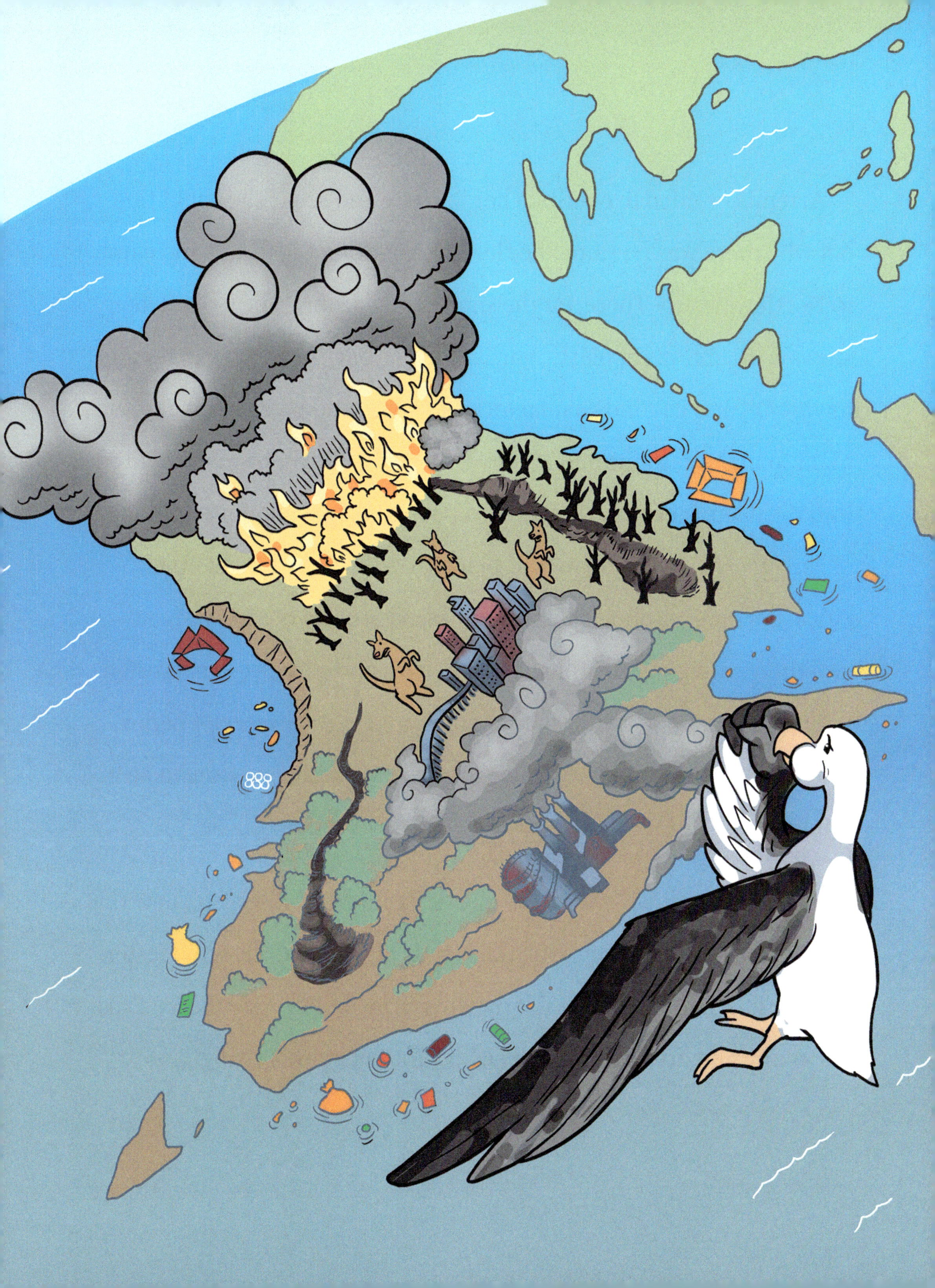

Artemis decided to continue her journey and see if she could find out why there was no rain. She flew over another big ocean and came to North America. There, on the shores, and around the Great Lakes, she saw mountains of plastic garbage in the bays near huge cities and towns. Oil and gas and coal factories burped, and belched all day and night to make gasoline and all kinds of plastic. Carbon smoke poured into the deep, blue sky. Thoughtless people rushed about in millions of cars, trucks, motor boats and planes, rushing to work, to other cities, burning gasoline all day. They raced around buying widgets and thingamijigs all the time. Their houses and garages were stuffed with gadgets and putputts. The children and adults were not happy. There was no time to play outside, no time to cuddle, no time to be in a muddle.

Artemis now knew where the plastic came from. She also knew that the carbon smoke was making the clouds bigger, and the wind faster. She felt it in her wings. She felt the air getting hotter, saw the ice caps shrinking and breaking apart quickly. The ocean was rising.

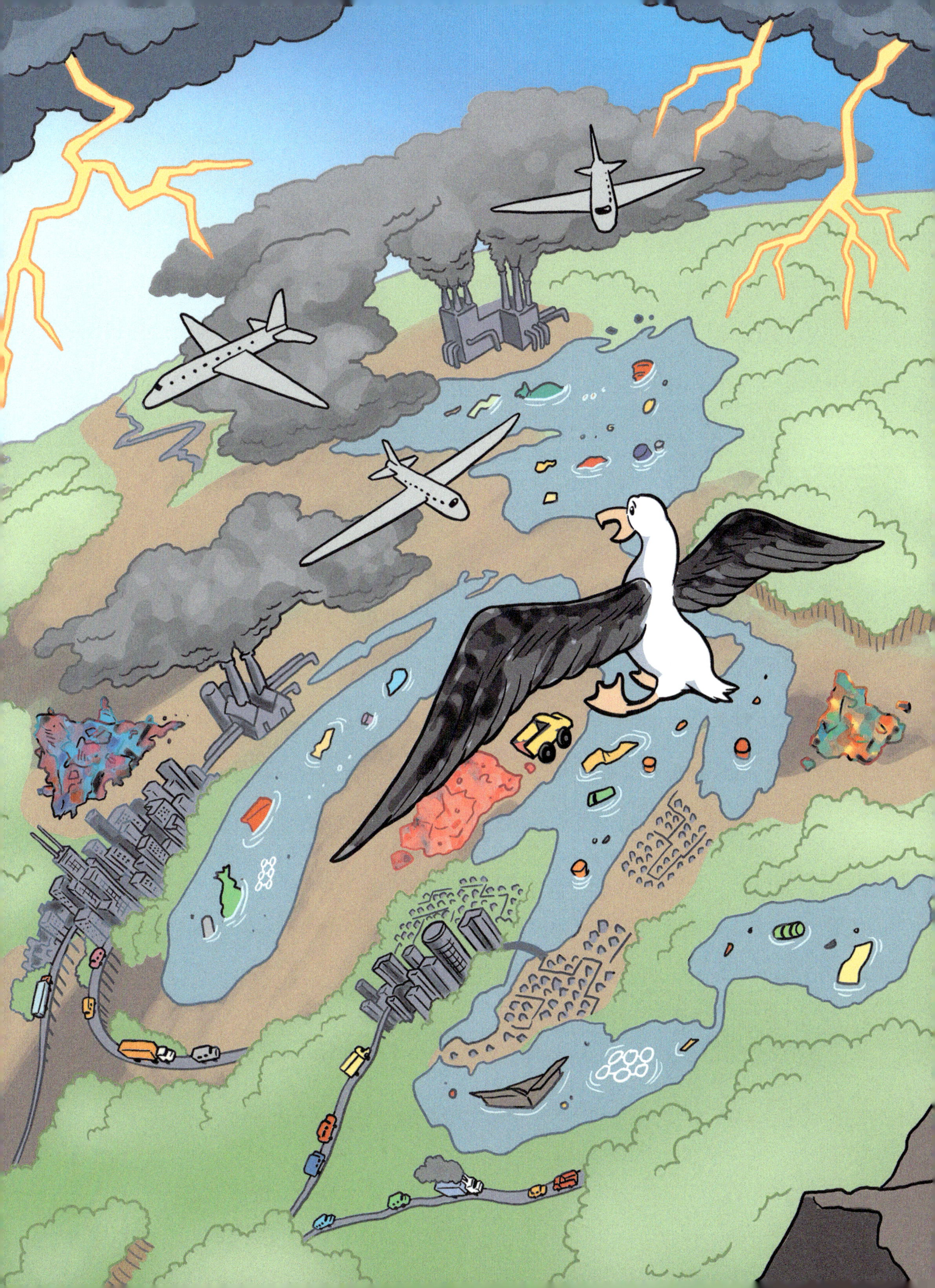

But the oil, gas, and coal company owners were greedy.

They had money in their shoes, money in their bulging wallets, money in their underpants, and money in their ears. But they still would not stop mining and selling oil, gas, and coal. They would not stop making plastics. They did not care that in some places there was no rain and in others way too much. They grubbed and lied and tricked the people to buy more and more. They never saw the sky. Their hearts were hard and cold.

Artemis was so sad to see the smoke turn into monster clouds, swirling and twirling in the wind. In some places, the rains thundered down in big curtains. Tornadoes crumbled factories, houses, and forests. The water in the oceans rose, the lakes and rivers rose and people stood on their toes, as it came up to their noses. Floods covered the fields, so no food could grow. The people cried, the animals asked why, and they tried to swim to hills and mountains to keep above the water.

At last the people were scared and shocked and said, "We can't breathe. We can't see. We can't live in this misery."

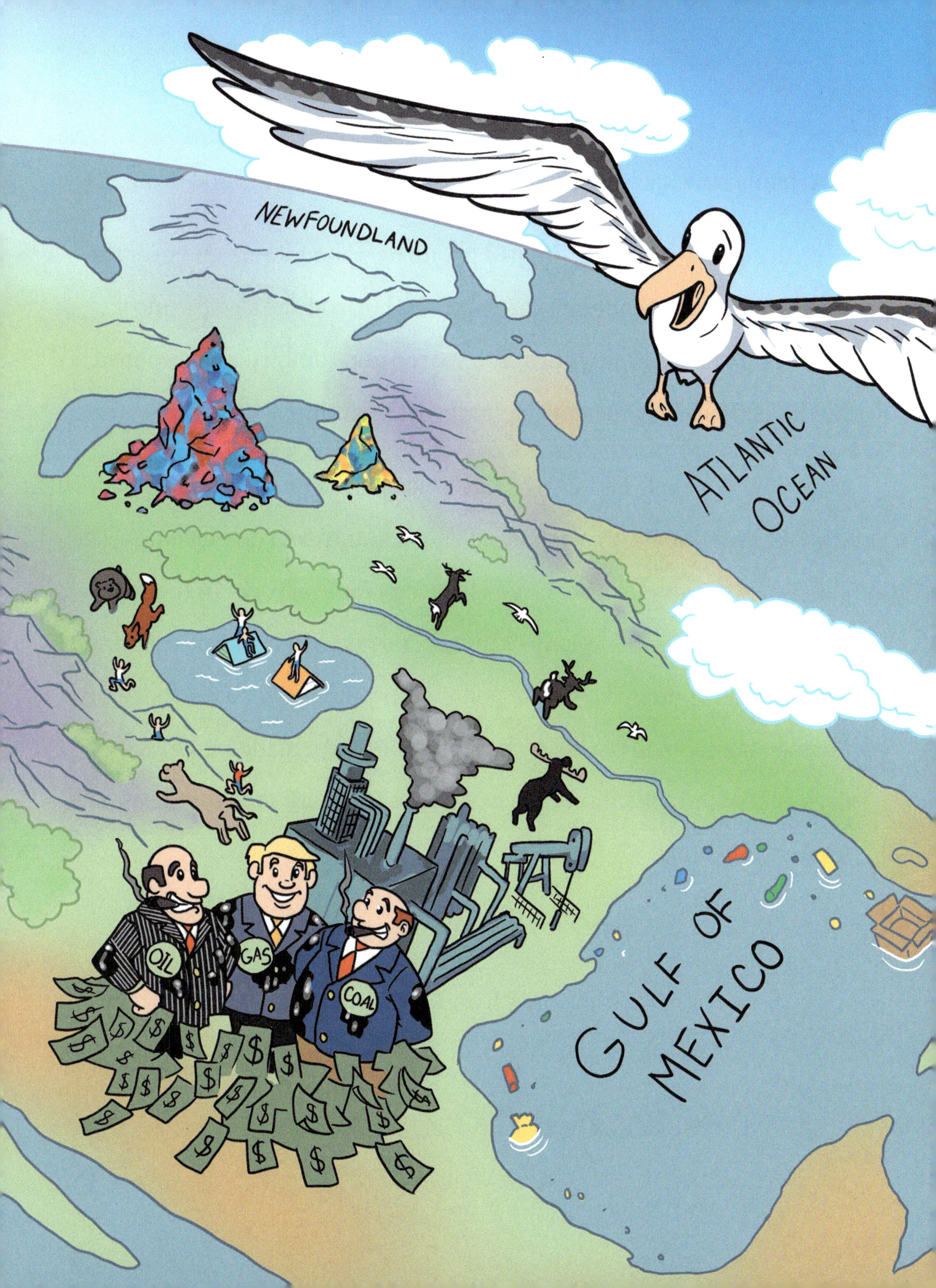
NEWFOUNDLAND
ATLANTIC OCEAN
GULF OF MEXICO
OIL
GAS
COAL

Artemis now knew what was causing all those storms and droughts.

With a heavy heart she flapped up, up over mountains and over the ocean to Europe. As the sun rose over the horizon she saw the most wonderful and exciting things. She soared and tumbled and swooped and looped, like a ballet dancer, her heart pounding with joy.

Everywhere, people had shiny, black solar panels on the roofs of schools, apartments, gyms, pools, factories, and even on airplanes. The panels use the sun to make clean electricity for everyone. In the bays there were wheels in motion – windmill farms with big arms catching wind to make power. No smoke polluted the air. No plastic garbage floated about. Small plastic garbage mounds were made into toys, decks and windmills. Everyone had lots of fun, with music, dances, and feasts. They sang:

> Earth my body Water my blood
>
> Air my breath and Fire my spirit
>
> We are one

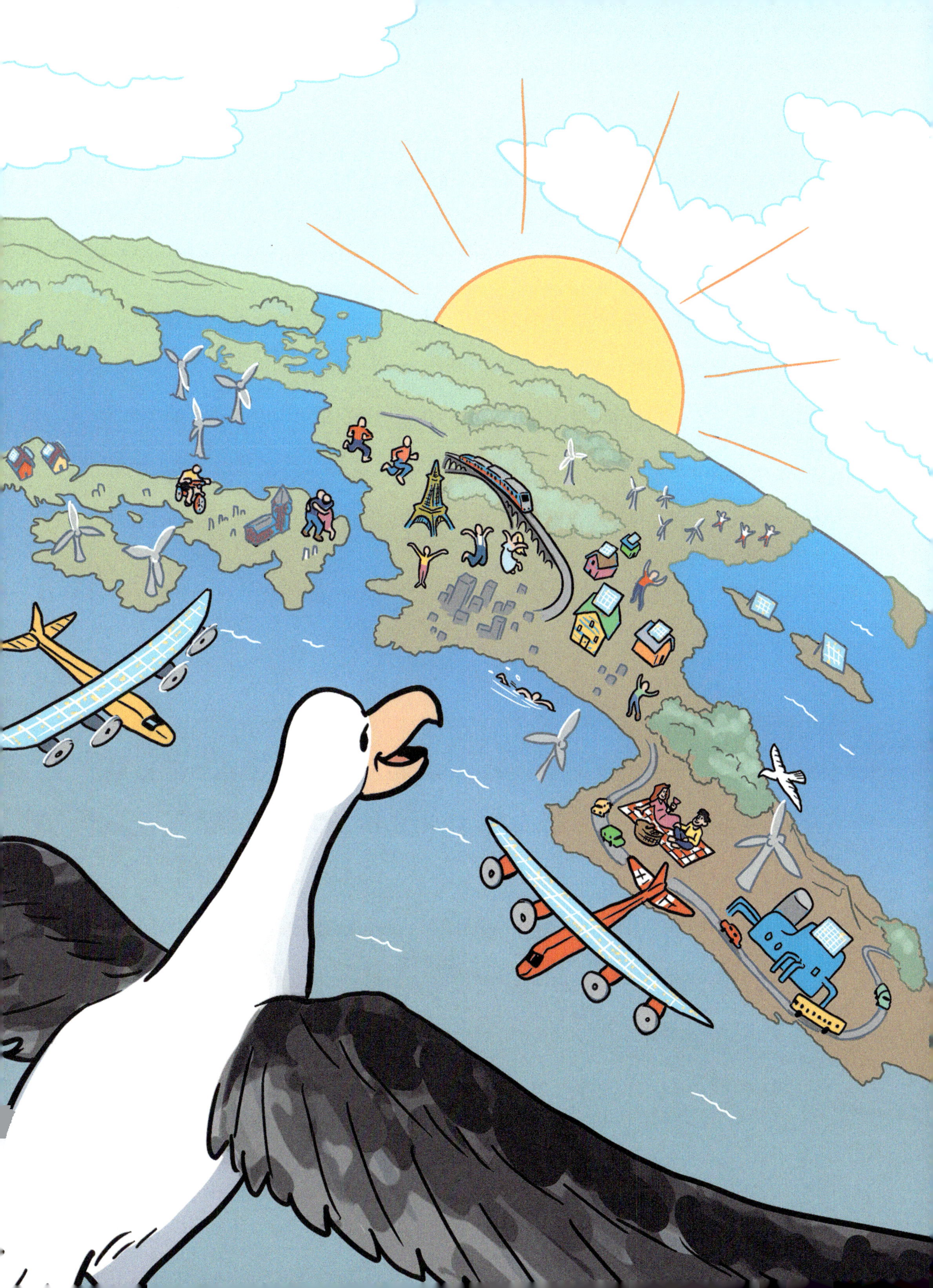

Excited, Artemis flies home carrying the good news and the bad. She sits down with her relatives and tells them stories of all she has seen. She has an idea. Working together, Artemis and the other albatrosses make drawings of solar panels and windmills, draw and write down how to make useful things from plastic garbage. The albatrosses are so giddy with joy they add instructions in all the languages of the world. They also include the words to their favourite, song to share with all the human beings, the animals and plants, and all birds of the world.

The next morning, hundreds of birds, backpacks full of packages, lift off the cliffs to fly in all directions of the globe.

The people love the surprise. Packages float down from the majestic, big birds. Children, their relatives and parents, all over the world go out onto the streets with placards, they write letters, email, texts and phone to demand changes to laws to stop making oil, gas and coal machines and help build cities with solar and wind and all kinds of renewable energy. The politicians listen and turn cities and towns into healthy places. The children and the big people hurry to change their oil, gas, and coal machines for clean solar, electric, and wind machines.

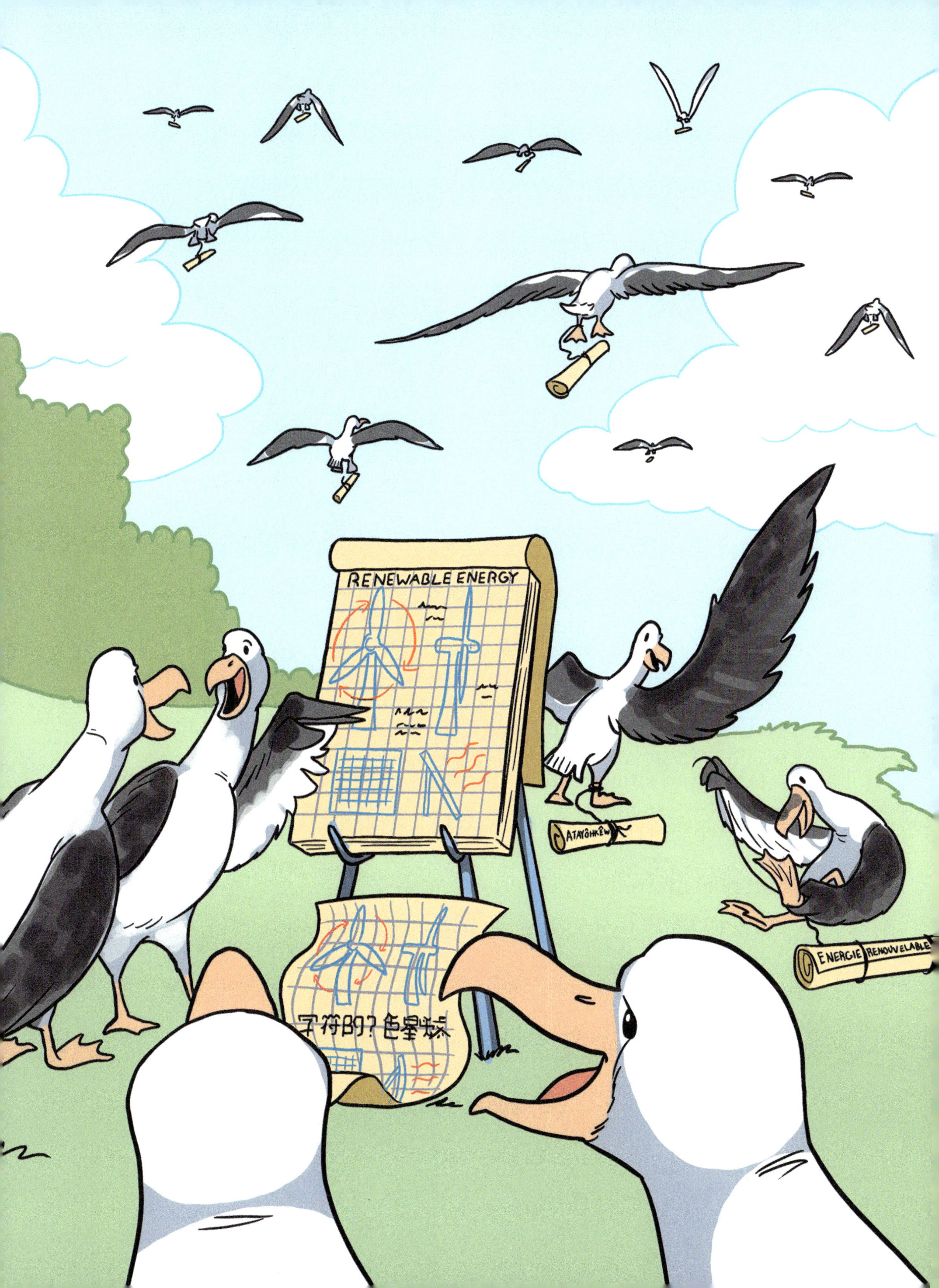
RENEWABLE ENERGY
ATAYÔHKÊW
ENERGIE RENOUVELABLE

Children, neighbours, and friends join together as planet partners. They build solar powered houses, solar powered factories, solar schools, and solar office buildings. Windmill farms make electricity from wind. Solar panels make electricity from the sun. The people grow food in gardens and public parks. Growing vegetables, fruits and grains turns carbon smoke into healthy things to eat. Some people go out to the oceans on big ships and drag in the terrible plastic islands bit by bit. They bring them home to recycle them into useful whatchamacallits. Artemis and the albatrosses dance on the cliffs. The oceans and the planet are becoming safe and beautiful again.

The rainy seasons are returning to Africa and Australia, so food and forests can grow. Rivers run fresh with fish and fresh water to drink. People dance and sing and feast together. Terrible tornadoes, droughts, and floods are less frequent. Ice-caps and glaciers are coming back. People all play and sing together:

> Earth my body Water my blood
>
> Air my breath and Fire my spirit
>
> We are one

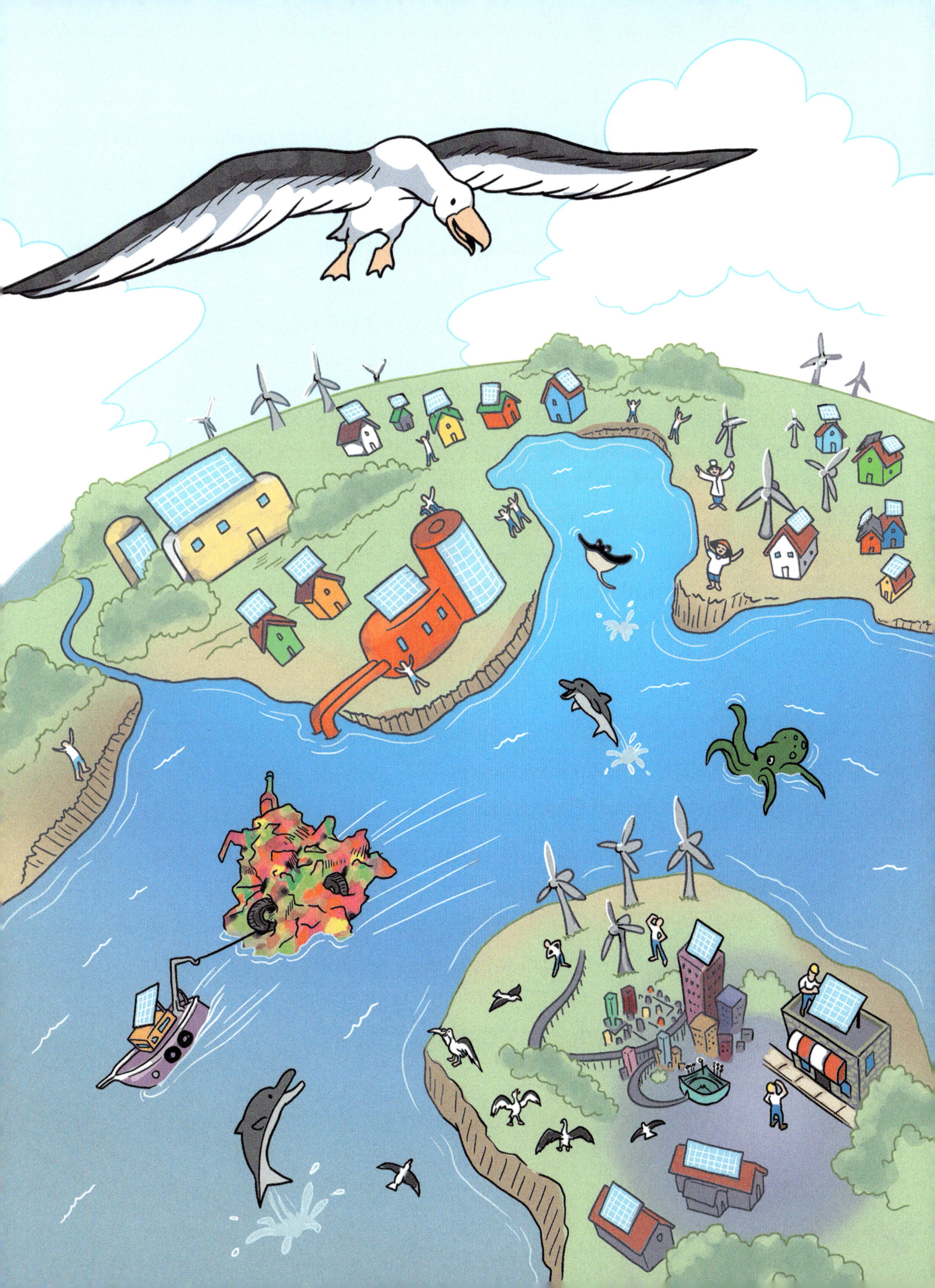

Glossary or Extra Information

Artemis: was the goddess of innocence, the hunt, the moon, and the natural environment. She was one of many goddesses of nature in ancient Greek mythology.

The albatross is the world's biggest bird that normally roams around the Pacific, Atlantic, and Indian oceans.

There are different kinds of albatrosses, but the biggest one has a wingspan of 3.5 metres. They can drink salty sea water without getting sick. They can fly 1,800 kilometres in 24 hours, and they spend many hours hunting for squid, crabs, lobsters, fish, and krill. They carry all this food in their gullets and fly all the way home to vomit their food down the throats of their hungry, wide-mouthed babies.

Climate Change means the changes of temperature on the planet caused by human beings burning oil, gas, and coal (fossil fuels) in industries, cars, power plants, and machines over the last 100 years. The carbon smoke is carbon dioxide which you often can't see, except at coal plants where it comes out as black smoke. It traps heat and is causing extreme changes in weather.

Albatross facts: http://a-z-animals.com/animals/albatross/, http://www.britannica.com/animal/albatross

Lesson Plans for all Grade Levels

http://www.climatechangelive.org/index.php?pid=180 includes
Renewable energy: http://pureenergies.com/ca/blog/top-10-countries-using-solar-power/

Plastic garbage in oceans: http://blogs.scientificamerican.com/observations/intolerable-beauty-plastic-garbage-kills-the-albatross/

Song Link "Earth My Body" - U-Tube

About the Author

Ingrid Alesich has lived in three countries: Egypt, Australia, and Canada. She now lives on the wide prairies of Saskatchewan in Canada. She has a bachelor of science degree in environmental sciences and ecology. She loves bird-watching, gardening, teaching, writing, and traveling to wonderful places on this beautiful planet.